Exploring **The**
BUILDING BLOCKS
of
Science

Book 3
STUDENT TEXTBOOK

REBECCA W. KELLER, PhD

REAL SCIENCE 4 Kids

Illustrations: Janet Moneymaker

Exploring the Building Blocks of Science Book 3 Student Textbook (softcover)

ISBN 978-1-941181-01-0

Published by Gravitas Publications, Inc.
Real Science-4-Kids®
www.realscience4kids.com
www.gravitaspublications.com

Contents

Introduction

Chemistry

Biology

Physics

Geology

Astronomy

Conclusion

Chapter 1 Using Science

1.1 Introduction

How do we use science to understand the world around us? How can science make our lives better? How have we improved the quality of the water we drink, the food we eat, and the air we breathe by using our understanding of how things work? What new technologies do we have that were created by using science?

Science isn't just about seeing how much an elephant weighs or measuring how fast an electron travels. Knowing facts is important, but using those facts to help make our lives better is what using science is all about.

1.2 Using Chemistry

Look around you. What are you wearing? What foods are you eating? Where are you sitting? What colors are you seeing on the pages of this book?

Every day you use things that were created or improved with chemistry. For example, many materials used for clothing have been created or improved by chemistry. Nylon and polyester were created by understanding how molecules work. Both nylon and polyester are found in many clothes and can make them stronger and longer lasting than cotton or wool.

YUM! I LOVE VANILLA!

Also, many modern foods contain chemicals that make them taste better. Vanilla is a favorite flavor additive for some foods, especially hot chocolate! Vanilla flavoring is made by a chemical process that separates the vanilla flavor from the vanilla bean.

Even the ink you see on these pages was created using chemistry! Blue, yellow, red, and black coloring were made by mixing different molecules. When these different colored inks were used in a printer, the words and images you see appeared on the pages of this book.

In Chapters 2-5 you will explore mixtures, molecular chains, and molecules in your body.

1.3 Using Biology

What did you eat for breakfast this morning? Did you grow it yourself, or did your mom or dad buy it at the store? When you have a tummy ache, do you drink peppermint tea to help it feel better? Do you have a large tree near your house to keep your yard shady and cool in the

afternoon? Every day you use things that were created or improved by an understanding of biology.

Today most food is produced using agriculture. Agriculture is a fancy word for farming—growing plants and raising animals. By understanding how plants and animals grow, what food they need, and how to keep them healthy, people can grow large amounts of food to nourish lots of people.

Biology also helps us feel better when we are sick. Many plants are medicinal, meaning they are medicine for your body. Peppermint tea, for example, is great for tummy aches. Aspirin is made from a chemical that was originally found in the bark of willow trees and is now created in the lab.

Knowing about biology also helps us create comfortable places to live. Landscape architects use plants and trees to make yards beautiful and functional.

Planting an herb garden is a great addition to a backyard, and having a tall tree with broad leaves can make a cool area of shade to play in during the hot summer!

In Chapters 6-9 you will learn about plants. You will learn how plants make and use food, the parts of a plant, and how plants grow and make more plants (reproduce).

1.4 Using Physics

How do you power a cell phone? How can you provide light at night so you can read when it's dark? How can you collect iron filings from the dirt in your backyard?

Every day you use and experience physics. Physics has given us an understanding of both simple and complex machines and everyday events. For example, physics has helped us develop small batteries to power technological devices such as cell phones.

Physics has also helped us understand electricity and how to move electricity from the power plant to your bedroom so you can read at night. Physics also helps us understand how magnets work and what metals might be attracted to them.

In Chapters 10-13 you will learn about the physics of atoms and molecules, electricity, and magnets.

1.5 Using Geology

When you go camping, do you put your tent on a grassy slope, or do you put it in a dry river bed? Can you grow a vegetable garden in the sand, or is it better to choose black soil? If you get lost in the forest, can you find your way home with a compass?

You and your ancestors have used geology to make decisions about where to travel, where to grow food and build houses, and how to navigate Earth's roadways and waterways.

For example, by studying how water flows and interacts with soils, geology helps us understand why placing a tent on a grassy slope is better than putting it in a dry river bed. A river bed could fill up with water if there were a sudden rain storm, but the rain would likely run off a grassy slope.

Understanding geology helps us grow food in the right places as we learn which rocks, minerals, and soils help plants grow and be healthy. By studying geology we can also learn about Earth's dynamics and what happens near volcanoes or in places where there are earthquakes.

In Chapters 14-17 you will learn about Earth's water system, how plants and animals interact with Earth, and how Earth's magnetic field helps us navigate.

1.6 Using Astronomy

Have you ever looked at the stars in the night sky? Do you know which points of light are stars and which are planets? If you don't have a compass, can you find your way home using the constellations?

The stars and planets in the night sky have been explored by people for many centuries. We use astronomy to help us learn about how the planets move, how the Sun produces heat and light, and where Earth is located in the cosmos. Astronomy helps us understand the world around us.

In Chapters 18-21 you will learn about our galaxy, other galaxies, comets, asteroids, and nebulae.

1.7 Summary

● Science can help us live better lives.

● Science helps us understand the world around us.

● We use science to create new technologies.

● What we learn from chemistry, biology, physics, geology, and astronomy helps us cure diseases, grow more food, make new inventions, and understand our world.

Chapter 2   Mixtures

2.1 Mixing

Have you ever put water and sand together in a pail? What did you get? A mud pie maybe!

Have you ever made a real pie, like lemon pie? If you have, you probably added eggs and flour, some table salt and oil, and maybe some water. What happened when you added all these things together? You probably mixed them with a spoon or a mixer.

In either case, what you ended up with is a mixture.

A mixture of sand and water or a mixture of eggs, oil, lemon, and water—both mud pies and lemon pies are mixtures.

2.2 Mixtures

You can make a mixture of blocks and rocks. You can make a mixture of rocks and sand. You can make a mixture of sugar and cinnamon and put it on your toast! All of these are called mixtures because all of these are made of more than one thing *mixed* together.

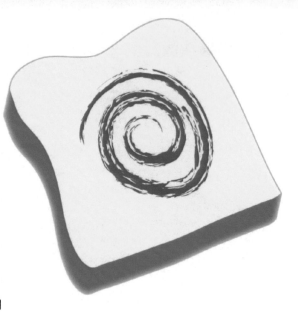

2.3 Some Mixtures Dissolve

Have you ever wondered why table salt disappears in water, but sand does not? Have you ever noticed that sugar disappears in water but not in oil or butter? When table salt or sugar disappear in water, we say they dissolve.

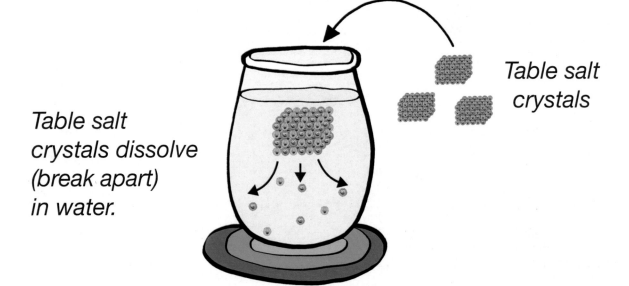

Table salt crystals dissolve (break apart) in water.

Table salt crystals

Some things will dissolve in water and some things will not dissolve. What makes some things dissolve and other things not dissolve?

2.4 Dissolving

As with everything else, it's the molecules in table salt and sugar that determine whether or not they will dissolve.

Molecules have to follow rules for dissolving or not dissolving, just like they have to follow rules for reacting or not reacting.

The main "rule" for dissolving is:

Like dissolves like.

This means, for example, that molecules that are "like" water *will* dissolve in water and molecules that are "not like" water *will not* dissolve in water.

This doesn't mean that the molecules have to be identical or *exactly* alike, they just need to have a few things in common.

Water

For example, what makes some molecules "like" water? Acid molecules have an H group (one hydrogen atom) and bases have an OH group (an oxygen atom and a hydrogen atom). If we look carefully at water, we see that it has BOTH an OH group and an H group! This is one of the things that makes water very special.

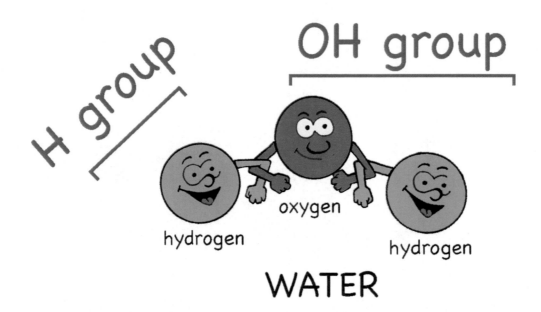

It is the OH group that makes molecules dissolve in water. Bases that have OH groups are "like" water and will dissolve in water. Other molecules, like alcohol, which is not a base but still has an OH group, will also dissolve in water.

Sugar is "like" water because sugar also has OH groups. Can you count how many OH groups sugar has?

1 Alcohol, sugar, and sodium hydroxide are "like" water—they have OH groups.

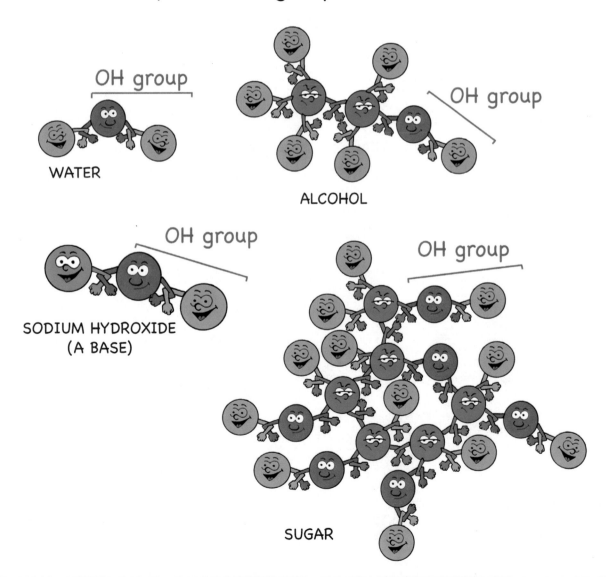

It's not just OH groups that make some things dissolve in water. For example, salt doesn't have OH groups like sugar, alcohol, and bases do, but salt dissolves in water. Salt dissolves in water because the water molecules break the salt molecules into pieces that mix with water.

2 Salt will dissolve in water.

First the salt breaks apart...

...and then the salt atoms mix
with the water molecules.

Oil, grease, and butter are not like water, so none of these will dissolve in water. Look carefully at the drawing that illustrates the type of molecule found in oil, grease, and butter. Can you tell why it is not like water?

3 Oil, grease, and butter are not like water.

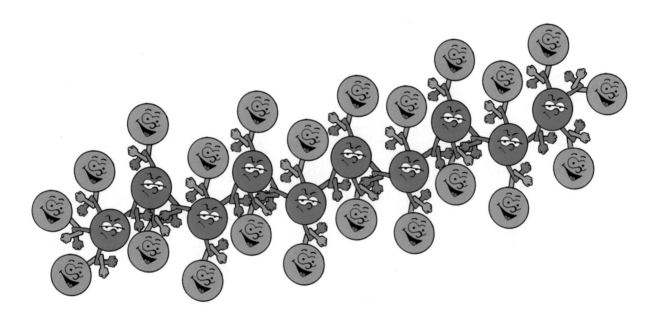

TYPE OF MOLECULE FOUND IN OIL, GREASE, AND BUTTER

2.5 Soap

Soap makes things like butter and grease dissolve in water. Soap can do this because the molecules that make up soap are a little like water and a little like oil.

4 Soap has an "oil-like" part and a "water-like" part.

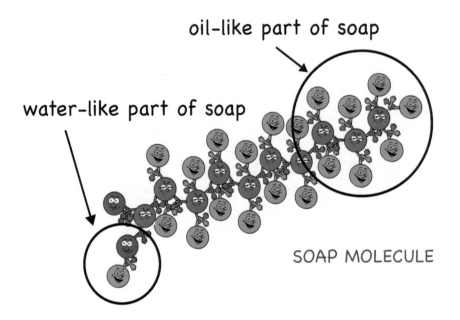

oil-like part of soap

water-like part of soap

SOAP MOLECULE

In a mixture of oil, soap, and water, the oily part of soap will dissolve in the oil, and the watery part of soap will dissolve in the water.

5 The "oil-like" part of soap dissolves in the oil, and the "water-like" part dissolves in the water.

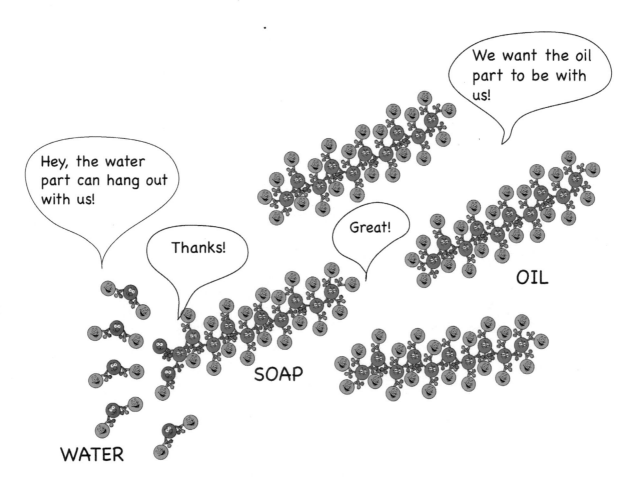

WATER, SOAP, AND OIL MIXTURE

Because the oil dissolves in the oily part of soap, and the watery part of soap dissolves in the water, a small droplet of oil and soap forms. In this way, the oil is "trapped" by the soap and water inside this little droplet.

This droplet can then be washed away by the water. This is how soap washes the grease off your hands!

6 Droplet of oil molecules and soap surrounded by water molecules.

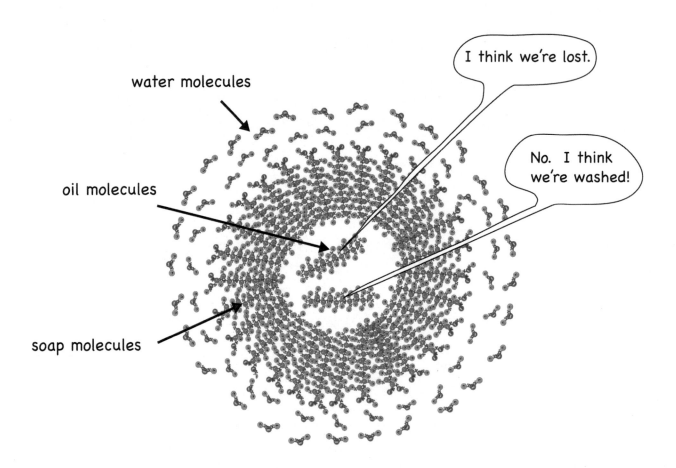

2.6 Summary

● Mud pies and lemon pies are mixtures.

● A mixture is anything that has more than two types of items in it.

● Some mixtures dissolve. Others do not.

● Dissolving depends on the kind of molecules in the mixture. Molecules that are "like" each other dissolve. Molecules that are "not like" each other will not dissolve.

● Soap is like both water and oil. This means that soap can make oil "dissolve" in water.

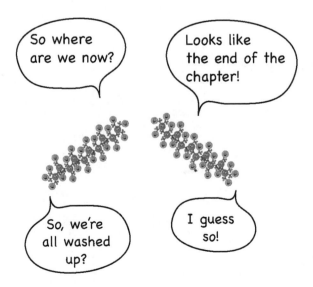

Chapter 3   Un-mixing

Chemistry

3.1 Un-mixing

In the last chapter we learned about mixtures, but how do we get things that are mixed to "un-mix?" Can we get the water and sand to "un-mix" from a mud pie? Can we get the eggs, sugar, water, and lemons to "un-mix" from a lemon pie?

Try to think of ways to "un-mix" a mud pie. What if you let the mud pie bake in the sun? What happens to the water? What happens to the sand?

3.2 Evaporation

You may know that the water "disappears" from the mud pie and the sand stays behind. We say that the water has evaporated. Evaporation is one way to "un-mix," or separate, mixtures that have water in them.

What happens if we leave the lemon pie to bake in the sun? Will the lemon pie "un-mix?" The water will evaporate, but what happens to the eggs, sugar, and lemons? They do not evaporate. In fact, they stay behind and we have a not-so-tasty lemon mess!

3.3 Sorting By Hand

Sometimes we can "un-mix" things, and sometimes we cannot. The mud pie can be "un-mixed," by the sun, but the lemon pie cannot.

Large things are usually easy to "un-mix." Even though when your mom tells you to clean your room, the large pile of toys may seem impossible to "un-mix"—with some work, it can be done.

All of the toys are easy to pick up because they are large. They can be picked up with your hands and put into separate bins or boxes.

3.4 Using Tools

What about a pile of sand? The sand cannot be easily picked up because each piece of sand is very small. It would take hours to pick up all of the sand by hand!

Fortunately a tool can be used. Can you think of a tool for picking up sand and "un-mixing" it from your carpet?

That's right—a vacuum cleaner! A vacuum cleaner can be used as a tool for "un-mixing."

In fact, tools are used all the time to "un-mix" things that are hard to "un-mix" with your hands. For example, sieves or colanders are used to separate hot spaghetti or hot potatoes from boiling water.

3.5 Using "Tricks"

There are other tools and other ways to "un-mix" mixtures of small things. What about molecules that you can't even see? Are there ways to separate molecules?

There are! In fact, scientists use a trick called chromatography to separate molecules. Using chromatography, you can un-mix many different kinds of molecules.

One type of chromatography is called paper chromatography. With paper chromatography, a piece of paper is used to separate small things like molecules. You can use paper chromatography to separate the small molecules that are in ink or dye, or even the molecules found in a colored flower!

Paper Chromatography

Ink colors "un-mixing" on the paper

Mixture of different colored ink

paper

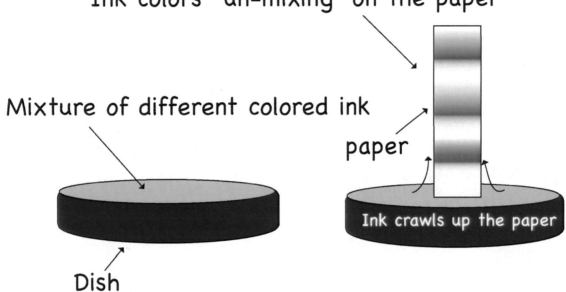

Ink crawls up the paper

Dish

3.6 Summary

- ○ The sand and water in a mud pie can be "un-mixed" by evaporating away the water.

- ○ Some mixtures, such as lemon pies, cannot be easily "un-mixed."

- ○ Mixtures of large things are easier to "un-mix" than mixtures of smaller things.

- ○ Tools, like vacuum cleaners and sieves, can be used to "un-mix" some mixtures.

- ○ A trick called chromatography can be used to separate molecules.

Chapter 4   Molecular Chains

Chemistry

4.1 Chains of Molecules

Sugar molecules hook together to form a long chain. These long chains of sugar molecules are called carbohydrates.

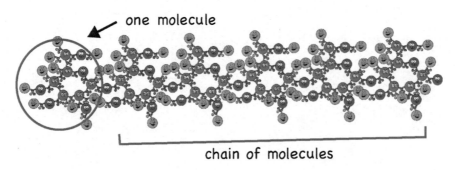

one molecule

chain of molecules

CARBOHYDRATE

There are other kinds of long chains not made of sugar molecules. Long chains can be made out of many different kinds of molecules. In general, long chains of molecules are called polymers.

Polymer

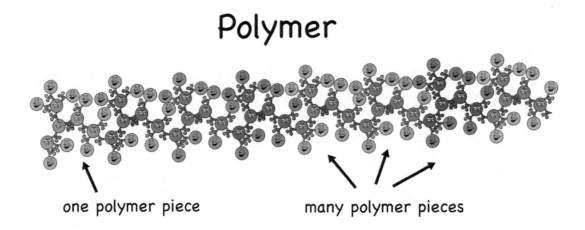

one polymer piece   many polymer pieces

Polymers are everywhere! Almost anywhere you look, you can find polymers. Your clothes, your toys, your food, and your hair are all made of polymers!

4.2 Different Polymers

Plastics are polymers.
Your toy car and parts of
your dad's car are made of
polymers.

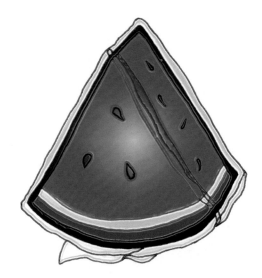

The plastic wrap you put
over your food is made of
polymers.

The plastic cup you drink from and
the plastic pen you write with are
made of polymers.

Rubber is also a polymer. The rubber hose outside in the garden, a rubber ball, and the rubber boots you wear on rainy days are all made of polymers.

Styrofoam is a polymer. The packing peanuts that come with your new chemistry kit are made of polymers. The Styrofoam cup that holds your dad's coffee or your lemonade is made of polymers.

Your clothing is also made of polymers. Cotton fibers, nylon fibers, polyester, and wool are all polymers. Polymers are everywhere!

4.3 Polymers Can Change

How can polymers make so many different things if they are all just long molecules?

Because different polymers have different properties (like being sticky or stiff), many different things can be made with polymers.

Sometimes long chains of polymers slide up and down next to each other. Polymers that are like this, such as glue and natural rubber, can be sticky.

Sometimes long chains of polymers are hooked together and can't slide up and down next to each other. Polymers like this can be hard or stiff and not sticky.

It is possible to change the properties of polymers by using chemicals or heat to cause a chemical reaction. For example, egg whites are made of polymers. When you cook an egg, you change the properties of the polymers inside the egg. The egg whites change from a clear, sticky liquid to a firm, white solid when you cook them. The polymers inside the egg whites change their properties because the heat causes a chemical reaction.

4.4 Summary

○ Polymers are long chains of smaller molecules that are hooked together.

○ Polymers are everywhere! Plastics, rubber, Styrofoam, and clothing are all made of polymers.

○ Different polymers give objects different properties. Some polymers make things soft, and some polymers make things hard or stiff.

○ You can change the properties of a polymer by use of chemical reactions or heat.

Chapter 5   Molecules in Your Body

Chemistry

5.1 Special Polymers

In the last chapter we looked at long chains of molecules called polymers. We learned that plastics, rubber, clothing, and food are all made of polymers.

Did you know that many of the molecules inside your body are also polymers? There are many different polymers inside your body. We will learn about two of them.

One very special kind of polymer is called a protein.

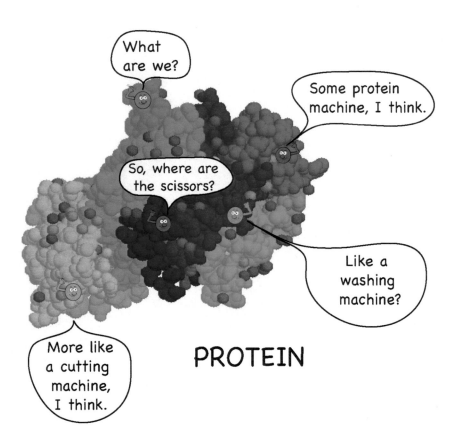

PROTEIN

A protein is a long chain of small molecules hooked together and often folded up into a special shape. It is the special shape of this folded chain of molecules that helps proteins do amazing things.

5.2 Proteins—Tiny Machines

Proteins are tiny machines inside your body that perform incredible tasks. In fact, proteins do almost all of the work inside your body.

Some proteins glue other molecules together.

Some proteins cut other molecules.

Some proteins copy other molecules and some proteins "read" other molecules.

Some proteins move other proteins or molecules from place to place.

Proteins do an amazing number of different jobs inside your body.

5.3 DNA—A Blueprint

One of the molecules that proteins read, cut, paste, and carry is called DNA. DNA is also a polymer.

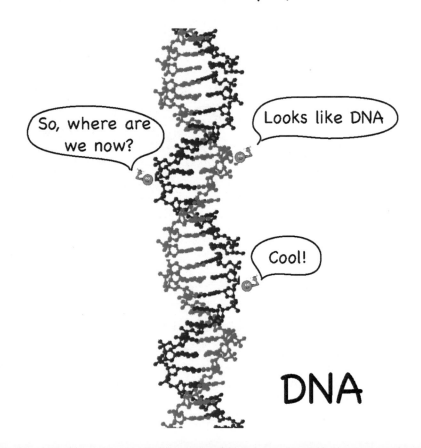

DNA

DNA is a very special molecule. It is not just any ordinary polymer. DNA is special because DNA carries your genetic code. A genetic code is like a set of instructions. Your genetic code determines if you will have brown hair or blonde hair. Your genetic code tells whether you will have green eyes or blue eyes, or whether you will have light skin or dark skin. The genetic code carried by your DNA is essentially the blueprint for your body.

Everyone has a different ar unique genetic code, or blueprint. You get your blueprint from your parents, and they got their blueprint from their parents. Your parent's parents got their blueprint from their parents, and so on. Your blueprint tells what you will look like and how tall you may grow, but your blueprint doesn't tell everything about you.

Where you live, what you eat, what you do, and even what you think make you unique and not like any other person who ever lived or who ever will live! Even identical twins who have identical DNA are different from each other. You are more than just your DNA. You are uniquely designed in every way.

5.4 Summary

○ There are polymers in our bodies.

○ Some polymers are called proteins. Proteins are tiny machines that glue, cut, copy, and carry molecules in your body.

○ Some polymers are called DNA. DNA carries the genetic code.

○ Your body is an amazing design of large and small molecules, polymers, and genetic information. You are uniquely designed.

Chapter 6 Plants

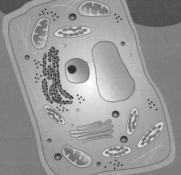

Biology

6.1 Introduction

When you go outside almost anywhere in the world, you will see plants. Plants live in deserts, near rivers and streams, on top of mountains, and even at the bottom of the sea. Plants give us food and medicine, materials for making homes and clothing, and most importantly, plants give us oxygen to breathe!

6.2 So Many Plants!

Plants are in the domain Eukarya. Because there are so many plants with different features, plants have their own kingdom. The Latin name for the plant kingdom is Plantae. There are over 400,000 different plants in the plant kingdom!

Some plants are very big, and some plants are very small. Some plants have flowers, and other plants don't have flowers. Some plants have long thin leaves, and some plants have wide leaves that spread out. Some plants have cones, and some plants have fruit. Some plants even have thorns.

Because there are so many different types of plants, scientists sort the plant kingdom into different divisions. A division is a group of plants that share similar features.

The plant kingdom is very complicated, and it can be difficult to determine exactly how to create divisions. Some groups of scientists divide plants into only a few divisions, and other groups of scientists divide plants into as many as 30 divisions.

6.3 Where Plants Live

Plants can live almost anywhere! Plants can live in a tropical rainforest, by the ocean, or on a frozen tundra. Plants can live where it is very cold, like regions by the North Pole, or where it is very warm, like the deserts of Arizona.

However, not all plants can live in all areas. Some plants need lots of water and can't live in a desert. Other plants need more sunlight and can't live where there are too many cloudy and rainy days. Many plants can't live at the top of mountains where there is little oxygen or soil, and only some plants can live in the salty ocean.

The area where a plant lives is called its environment. Environment refers to the conditions surrounding a plant where it lives, including features such as temperature, amount of water, amount of sunlight, and type of soil.

Although plants can live in different places, most plants need some sunlight, some water, and temperatures that are neither too hot nor too cold. Plants can't live in outer space. As far as we know, no other planet in our solar system has the right environment for plants to grow.

6.4 Plant Cells

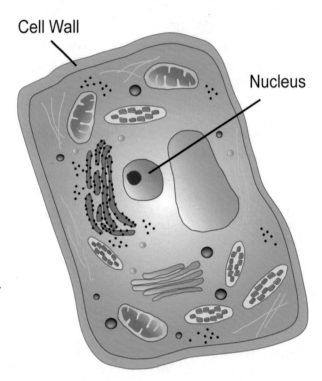

Cell Wall

Nucleus

Plant cells are different from cells found in bacteria and archaea. Plant and animals cells are called eukaryotic cells, and unlike bacteria or archaea, eukaryot cells have a nucleus which contains the DNA.

Plant cells are also different from animal cells because plant cells have a cell wall. Cells walls make plant cells rigid and give plants the strength and ability to stand upright and to bend but not break in the wind.

6.5 Summary

- Plants are in the domain Eukarya and the kingdom Plantae.

- The plant kingdom is sorted into different divisions. A division is a group of plants that share similar features.

- The area surrounding a plant where it lives is called its environment.

- Plant cells have a nucleus and a cell wall.

Chapter 7 Food for Plants

Biology

7.1 Introduction

How do plants eat? Do they eat spaghetti or french fries? Have you ever met a plant at the diner drinking a milk shake?

No, probably not. Plants can't eat spaghetti or french fries or drink milk shakes.

In fact, most plants have to make their own food. They make their food using sunlight.

7.2 Factories

Cells are like little cities. Just like cities have factories that make bread or cereal, cells also have places inside them that act like factories.

We call these little factories organelles. An organelle is like a little organ inside cells that does a special job.

There are many different kinds of organelles inside cells. To make food, a plant cell has an organelle called a chloroplast.

A food factory (called a chloroplast) is a kind of organelle.

7.3 How Plants Make Food

Plants use sunlight to make food. This process is called photosynthesis. Photo means "light" and synthesis means "to make." So photosynthesis means "to make with light."

Plants do this by catching the light from the Sun in a very special molecule inside their cells. This molecule is called chlorophyll. The job of the chlorophyll molecule is to catch sunlight and use it to make sugar!

Part of the chloroplast that catches sunlight (chlorphyll molecule)

FOOD FACTORY

Green parts of a plant have food factories (chloroplasts)

leaves

stems

7.4 Food Factories

Food for a plant is made in all of the green parts of the plant. This is because the chloroplasts are found in all these green parts. Green plant leaves and green plant stems both have chloroplasts.

The green color in a chloroplast comes from the chlorophyll molecule. The chlorophyll molecule catches the sunlight and also makes the leaf or stem green.

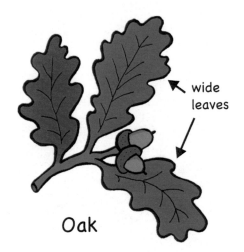
wide leaves

Oak

7.5 Different Leaves

There are many different kinds of leaves on plants and trees. Some trees, like oak trees, have very wide leaves. This is so they can collect as much sunlight as possible.

Other trees, like a willow tree, have lots of narrow leaves. Having lots of leaves will also help a tree collect as much sunlight as possible.

narrow leaves

Black Willow

What happens to leaves when they change color? Why do leaves turn yellow in autumn and then fall off in winter? Leaves change color in autumn because in autumn the days are shorter.

It is a lot of work for a tree to use sunlight to make sugar to use for food. When there isn't enough sunlight to do this, the tree gets rid of its leaves. A tree knows when the days are getting shorter, and this causes the green chlorophyll molecules to go away, the tree to stop making food, and the leaves to first turn color and then fall off.

7.6 Summary

● Most plants make their own food by photosynthesis.

● Plants have tiny factories called chloroplasts inside their cells. Chloroplasts are used by plants for making food.

● The green parts of a plant, like leaves and stems, contain chloroplasts and make the food for the plant.

● Leaves change color in autumn when the tree doesn't get enough sunlight to make food.

Chapter 8 Plant Parts

8.1 Introduction

Many plants live and grow in soil and above the soil—in the air. The part of the plant that lives in the soil is the root. The parts of the plant that live in the air are the leaf, stem, and flower. These different parts of a plant are called organs.

Plants get what they need to live and grow from the ground (the soil), and from above the ground (sunlight and air).

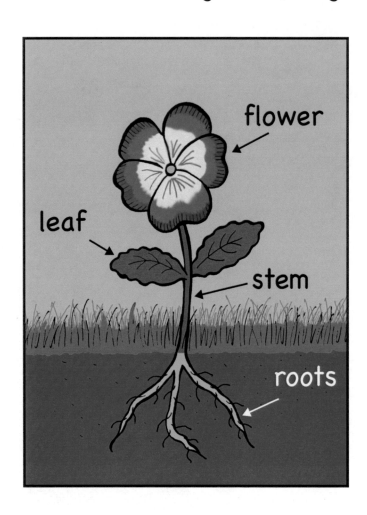

8.2 In the Ground: The Roots

Plants cannot walk around or fly to go get their food, like insects and other animals do. Instead, plants have roots in the dirt. The roots take nutrients from the soil, and the plant uses these nutrients to make food. The soil also holds the water that plants need in order to grow.

Roots also help keep plants from getting tossed around by wind or bad weather. Plants help keep the soil from blowing away.

8.3 Above the Soil: Leaves, Stems, Flowers

Above the soil, we find the leaves and stem. We saw in Chapter 7 that the leaves and stem of a plant take in the Sun's energy to make food. The Sun gives plants energy to make food.

For flowering plants, the flower is also above the soil. The flower is the part of the plant that turns into the fruit. The fruit holds seeds for new plants.

Plant plankton
Photo Credit: Chris Parks/imagequestmarine.com

8.4 Other Places Plants Live

Plants can also live in oceans. There are different kinds of plants that live in oceans. Some plants are very tiny, like plant plankton. Plant plankton provide food for animals in the ocean. Most plant plankton are microscopic, meaning you can only see them with a microscope.

Other ocean plants can be very large, like kelp. Kelp is a type of seaweed and can be several yards long. Seaweed can be red, brown, or green.

Like land plants, ocean plants need sunlight to make food. So most ocean plants are found close enough to the surface that they can get sunlight.

8.5 Summary

- Plants have different organs that help them live and grow.

- Below the ground, the roots help the plant get nutrients and water from the soil to make food.

- Above the soil, the leaves and stem help the plant take in sunlight to make food.

- The flowers are also above the soil. The flowers turn into fruit which hold the seeds for new plants.

Chapter 9 Growing a Plant

Biology

9.1 The Beginning: Seeds

Plants grow from seeds. Seeds can be lots of different shapes and sizes. A seed can be tiny or large, blue or brown, round or skinny.

Inside a seed there is a tiny plant called an embryo.

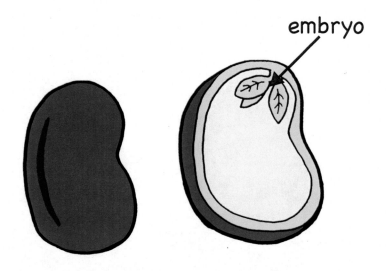

If a seed gets planted in the soil and if there is enough water, the embryo will start to grow. The water will make the inside of the seed swell, and the little embryo will break out of the seed and grow into a baby plant.

9.2 The Middle: Baby Plants

Once the embryo breaks out of the seed, it starts to grow. As it grows, it will change from an embryo to a baby plant. A baby plant is called a seedling.

As the seedling grows, it will push its roots down into the soil so that it can get water. It will also start to grow leaves so that it can get sunlight to make food. The sunlight will help the seedling straighten out and grow into a big plant.

9.3 The Finish: Flowers and Fruit

Once the plant is fully grown, it will be able to make baby plants. There are different kinds of plants and different ways plants make baby plants.

One kind of plant is called a flowering plant. A flowering plant produces flowers. The flowers produce fruit, and the fruit hold the seeds.

9.4 Starting Again: The Life Cycle

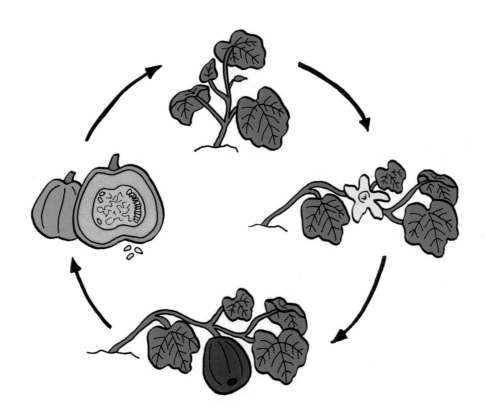

When the baby plant has grown into a big plant, and when the big plant grows the flowers, and when a flower makes the fruit, and when the fruit drops the seed—a new baby plant can grow. This is called a life cycle. A cycle is something that repeats itself. So a life cycle is how life repeats itself.

It takes a flowering plant to make a seed of a flowering plant. And it takes a seed of a flowering plant to make a flowering plant. This is how flowering plants make new flowering plants.

9.5 Summary

- Plants grow from seeds.

- Inside a seed is a tiny plant called an embryo.

- A baby plant is called a seedling.

- A flowering plant produces flowers that make fruit that hold the seeds for new plants.

Chapter 10   Energy of Atoms and Molecules

Physics

10.1 Atoms and Energy

Every material object has mass. Simply put, mass is how heavy something is. A tree has mass. A football has mass. You have mass.

Inertia is a force that keeps objects still and keeps objects moving. The more mass an object has, the more inertia it has. That is, the more mass an object has, the harder it is to start the object moving or to get the object to stop moving.

Every material object has mass because every object is made of atoms. Atoms are what make objects into objects. A tree is made of atoms. A fish is made of atoms. Your body is made of atoms.

oxygen atom

hydrogen atom

phosphorus atom

Phosphoric Acid
A molecule

When two or more atoms are connected together, a molecule is formed. Phosphoric acid is a molecule made of the atoms oxygen, hydrogen, and phosphorus.

When molecules interact with other molecules or atoms, a chemical reaction takes place.

What happens when you add baking soda to vinegar? Bubbles start to form, and the baking soda turns to foam. This is a chemical reaction. It takes energy for a chemical reaction to take place.

Where did the baking soda and vinegar get the energy? The energy is inside the baking soda and vinegar molecules. This is called chemical energy. Atoms and molecules have stored chemical energy. When a chemical reaction occurs, the energy stored inside the molecules is released and used to do work. Work is done when a force moves an object or changes its shape.

You can see the work being done if you put baking soda and vinegar together in a bottle and close the lid. The stored chemical energy of the baking soda and vinegar in the bottle will release, causing a chemical reaction.

The chemical reaction will make gas which will begin to press on the sides and top of the bottle and will put pressure on the lid. If there is enough energy used and enough gas created, the bottle might explode and the lid pop off! This is stored chemical energy being used to do work.

10.2 Energy for Cars

How does your car get energy for moving? Does your mom feed the car hamburgers? Or do your parents take it to a field and let it graze on the grass with the cows?

No! Your parents take your car to the gas station to fill the gas tank with gasoline.

Gasoline is a type of stored chemical energy used for fuel by cars, motorcycles, airplanes, and boats. The stored chemical energy in gasoline is released when your mom or dad starts the car. When the chemical energy is released, it can be used to move parts of the car so that the car's wheels can roll down the street.

10.3 Energy in Food

The cereal you eat for food is a type of stored energy. Foods, like potatoes, cereal, and bread, contain a type of stored chemical energy called carbohydrates.

Carbohydrates are special kinds of molecules that living things use for fuel. Sugar is a type of carbohydrate. Our bodies use lots of carbohydrates for energy.

The body breaks down carbohydrate molecules and uses the energy to move muscles, to make our heart beat, and even to think! It takes lots of energy to think, and eating carbohydrates is one way to give our bodies the fuel needed for thinking!

10.4 Batteries

Batteries are another type of stored chemical energy. What happens when you put batteries in a flashlight or in a game player?

Inside a battery are metals and chemicals. When the metals and chemicals come in contact with each other, a chemical reaction occurs. The chemical energy inside the battery is changed into electrical energy, and the electrical energy runs the flashlight or game player. We'll learn more about electrical energy in Chapter 11.

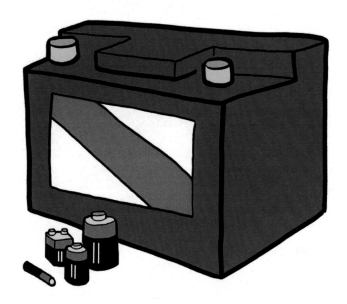

10.5 Summary

- Chemical energy comes from atoms and molecules and is released in chemical reactions.

- Gasoline is a type of stored chemical energy used by cars, boats, and motorcycles.

- Carbohydrates are a type of stored chemical energy in food and are used by living things for fuel.

- Batteries are a type of stored chemical energy used by flashlights and game players.

Chapter 11 Electricity

Physics

11.1 Introduction

In Chapter 10 you learned about chemical energy. Recall that chemical energy is energy inside atoms and molecules. When a chemical reaction occurs, chemical energy is released from the atoms and molecules.

You also learned that sometimes chemical energy is converted into electrical energy. Batteries convert chemical energy into electrical energy. But exactly where does the electrical energy come from?

11.2 Electrons

Electricity, or electrical energy, comes from electrons. An electron is part of an atom.

Look at your body. Notice that your body has different parts. You have eyes for seeing, legs for walking, and arms for picking up objects. Inside your body you also have lots of parts. You have lungs for breathing, a stomach for digesting food, and a heart for pumping blood. You have many parts to your body, and each part does something different.

In a similar way, an atom has different parts. An atom has three main parts called protons, neutrons, and electrons. The protons and neutrons are in the center of the atom, and the electrons move around outside this center.

The electrons are what make atoms stick to each other to form molecules during a chemical reaction. In the following illustration of a carbon atom, the "arms" represent the electrons that are used by carbon to stick to other atoms. The yellow and blue balls inside the carbon atom represent the protons and neutrons

The electrons of an atom can jump back and forth during a chemical reaction. That is, atoms can exchange electrons.

You can't give your arms to someone else when you meet them, but an atom can give part of itself (the electron) to another atom to form a chemical bond. It is the moving of these electrons that causes electrical energy. In metals, electrons jump from atom to atom all the time. Some metals have lots of electrons that jump from atom to atom.

11.3 Electrons and Charge

What happens when you rub a balloon in your mom's hair? If you pull the balloon away just slightly, your mom's hair will travel with the balloon. Physicists say that the balloon is charged. In this case, "charged" means that the balloon is attracted to the hair and the hair is

attracted to the balloon. But why? Why does the balloon attract your mom's hair and the hair attract the balloon? What makes the balloon "charged?"

Electrons make the balloon charged. Electrons have a charge. That means that they have the ability to attract or repel other things that also have a charge. Protons have a charge but neutrons are neutral. Neutrons have no charge (that is why they are called neutrons, because they are neutral).

Physicists and chemists say that electrons are negatively charged and protons are positively charged. This just one way to say that their charges are opposite. This is a "rule" that scientists follow so it is easier to talk about charges.

In physics, opposites attract. Positive charges attract negative charges, and negative charges attract positive charges. Therefore, protons attract electrons, and electrons attract protons because they have opposite charges.

When you rub a balloon in your mom's hair, electrons hop from your mom's hair to the balloon. Because there are now more electrons on the balloon and electrons have a negative charge, the balloon is negatively charged. Your mom's hair lost some electrons and now has more protons that have a positive charge, so her hair is positively charged. Because your mom's hair and the balloon are charged with opposite charges, they attract each other!

11.4 Electrons and Force

Forces cause things to move. This is true for large things, like wagons, and small things, like electrons. Electrical forces cause electrons to move.

Have you ever rubbed your feet on the carpet and then touched a door knob? If the air was dry enough, you probably felt the effects of electrical forces moving

electrons. You might have felt a small shock when —ZAP!— you touched the door knob. This is electrical force moving electrons.

11.5 Summary

● Electricity, or electrical energy, comes from electrons.

● Electrons are part of an atom. Atoms have protons, neutrons, and electrons.

● Electrons can move from atom to atom.

● Electrons are negatively charged. Protons are positively charged, and neutrons are neutral.

Chapter 12 Moving Electrons

Physics

12.1 Introduction

In Chapter 11 you learned how electrical energy comes from the movement of electrons. You learned that electrons can jump from atom to atom, or even from hair to balloons! You also learned that electrons are charged and that scientists say that electrons have a negative charge and protons have a positive charge.

Everything that is made of atoms has electrons. All material objects have electrons. Wooden tables have electrons. Marshmallows have electrons. Frogs have electrons.

Many kinds of materials will allow a small number of electrons to move through them. But only some types of materials allow lots of electrons to move through them. Scientists say that these types of materials conduct electricity. This means that lots of electrons can flow through them, like water flowing through a garden hose.

12.2 Electrons in Metals

If you look at a toaster, you notice that it is connected to a cable, and this cable is plugged into the wall. Inside the wall there is another cable that goes outside and connects to a big pole. This pole has cables on it that bring electricity into your house. In

some houses the cable in the wall goes to electrical cables that are buried under the ground instead of being on poles. In either case, the toaster needs to be plugged into the wall to get electricity in order to work.

This is also true for your computer (if it doesn't run on a battery). It is also true for your television set or video game player. All of these items need electrical energy to work. And all of these items are connected to wires or cables that are plugged into the cable in the wall that brings the electricity into your house.

If you open up one of the cables, you might see that it is made of metal wires. Copper metal is often at the center of most cables or wires because copper can conduct electricity. The reason copper can conduct electricity (allow electrons to flow) is because some of the electrons on the copper atoms are very loosely attached. That is, the electrons can easily jump from one atom to the next.

Copper atoms have lots of electrons. In the following illustration, the "arms" on each copper atom are replaced with dots to represent the electrons.

electrons as arms replaced with dots

In a piece of copper metal, lots of copper atoms are next to each other. This means that there are lots of electrons free to hop around. This is why copper is a good conductor.

Just like water can flow through a garden hose because the water molecules are not attached to the hose, electrons can flow through a metal wire because the electrons are not tightly attached to the metal atoms.

12.3 Electrons in Other Materials

You saw in the introduction that all materials have electrons. Marshmallows, teddy bears, wooden tables, and popcorn all have electrons. But these materials are not used for moving electrons. Scientists say that these materials do not conduct electricity. That is, they do not allow the electrons to hop from one atom to the next.

The electrons in materials that do not conduct electricity are more tightly held to the atoms. Because they can't hop from atom to atom, there is no flow of electrons through the materials.

Scientists call these materials insulators. If you look closely at the cable that connects your toaster to the wall, you will see that it is covered in plastic. Plastic is an insulator. Plastic does not allow electrons to move through it, so it does not conduct electricity.

Insulators keep the electrons from moving from wires to your hands. This is important because even though you are not made of metal, your body will conduct electricity! If you touch a wire used for moving electricity and it does not have a plastic covering, you could get a big shock! This is why it is always important to be careful not to touch electric wires or electric outlets. The amount of electricity that moves through the wires is too large for your body and can hurt you.

12.4 Summary

- Every material substance (everything that is made of atoms) has electrons.

- Metals are called conductors.

- Electrons move from atom to atom in a conductor much like water moves through a garden hose.

- Other materials, like plastic, are called insulators. Electrons do not move through these.

Chapter 13 Magnets

Physics

13.1 Introduction

In Chapter 11 you learned that electrons have a negative charge and protons have a positive charge. Remember, this is just one way to say that the charges are opposite.

If you look around, you can see lots of different kinds of opposites. Black is the opposite of white. Wet is the opposite of dry. Dark is the opposite of light. North is the opposite of south. East is the opposite of west.

Sometimes opposites attract. In Chapter 11 you learned that positive charges attract negative charges—opposite charges will attract each other. This attraction creates the force that holds atoms together.

Some materials create attractive forces that aren't charged. A magnet is a type of material that will create an attractive force, but a magnet is not charged. Although a magnet is not charged, a magnet has opposite poles, and the opposite poles attract.

13.2 Magnetic Poles

All materials not only have electrons, but all the electrons are spinning. Magnets are usually made of nickel or iron. Some materials, like copper, don't make magnets. In metals that aren't magnetic, there are an equal number of electrons spinning. But in metals that are magnetic, like nickel or iron, there are an unequal number of electrons spinning. Because these metals have an unequal number of electrons spinning, they create magnetic poles.

One way to think about magnetic poles is to imagine a box full of marbles. Imagine that you have an equal number of marbles. Imagine also that each marble is half white and half black. To make it simple, imagine that you also gave all the marbles a "rule."

The rule is: "The marbles have to be balanced." So, for every marble with the black side facing forward, there must also be a marble with the black side facing backward. For every marble with the black side  facing upward, there must be a marble with the black side facing downward. This way the black and white colors on the marbles are balanced.

If you throw all the marbles into the box, there will be a mixture of marbles. Some of the marbles will have the black side facing up; some will have the black side facing down. Some marbles will have the black side facing forward and some backward. The marbles will be mixed, but because there is an equal number of marbles, all the directions the marbles are facing balance out.

Now imagine that you throw one more marble in the box. This marble has the black side facing up. But there isn't another marble to balance this one out. So what do the marbles do?

In a metal, this is what happens with the electrons. Because there is an extra electron on metal atoms, the spins are not balanced.

One way to get more balance is for all the marbles to line up in one direction. When the marbles do this, the effect of one extra marble is not so noticeable.

All the black sides face one way, and all the white sides face the other way. Since all the marbles are facing the same way, you could say that one side of the box is "white" and the other side is "black." In this way, the box has "opposite" sides."

In a metal, when the extra electron gets all the electrons spinning in the same direction, it is just like having all the marbles line up with the white sides pointing in one direction and the black sides pointing in the other direction. In a metal, this creates a magnetic pole.

Because the poles in a magnet are not charged, we don't call them "positive" and "negative." Instead we say "north" and "south." The north pole and the south pole are opposite and attract each other.

Sometimes magnets will have the letters "N" and "S" written on them. These letters mean "North" and "South."

13.3 Magnets and Force

If you have played with magnets, you may have observed how one pole of a magnet will strongly attach to the opposite pole of another magnet. You also might have observed how the same poles of two magnets won't go together. No matter how hard you try to push them together, the same poles will not touch. If you hold the magnets in your hands, you can feel the force of the poles attracting each other or pushing each other away. This force is called a magnetic force.

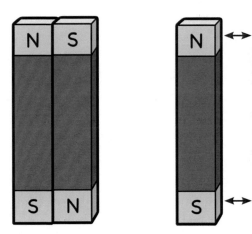

Magnetic forces are caused by the spinning electrons inside magnetic metals. The forces generated by magnets can be quite strong. Magnets can be used to lift cars or hold heavy equipment together. The Earth is also a huge magnet! If you travel north anywhere on the Earth, you'll end up at the North Pole. If you travel south anywhere on the Earth you'll end up at the South Pole. The North and South Poles are the ends of the magnet we live on called Earth!

13.4 Summary

- Only some metals, like nickel and iron, can be magnets.

- A magnet has opposite poles called a north pole and a south pole.

- The poles of a magnet are not charged.

- Magnetic force is caused by spinning electrons.

Chapter 14   Our Water

Geology

14.1 Introduction

Do you ever play in the rain and wonder how the water got into the clouds? Have you watched water flowing in a river and wondered where it comes from and where it goes? Have you ever noticed that ocean water is salty and lake water is not and then wondered why this is so?

Water is very important for life on Earth. Without water, life could not exist. Geologists study water and how it travels around the Earth.

14.2 Hydrosphere

The hydrosphere is the name for the water part of Earth. All the water on Earth makes up the hydrosphere. The hydrosphere includes all the water in lakes, rivers, and the oceans. It also includes rain, ice, snow, and the water in clouds and in the ground.

Water exists in three forms—as a liquid (flowing water), as a solid (ice and snow), and as water vapor (in the clouds). Part of the way water moves around the Earth is by changing from one form to another. Liquid water in oceans, lakes, and rivers evaporates, or changes from its liquid form to water vapor which is water's gaseous form.

When liquid water freezes to become ice and snow, the water changes to its solid form. When ice and snow melt, water returns to its liquid form.

14.3 The Water Cycle

The way water moves around Earth is called a cycle. Recall that a cycle is a series of events that repeat. We can think of the water cycle as beginning when liquid water flows from rivers into the oceans and then evaporates. Evaporation puts water into the atmosphere where it forms clouds. Then rain on the land puts the water back into the rivers. The cycle begins over again when this river water flows into the oceans.

14.4 Earth Is a Water Planet

How much water is on the Earth? If you look at a globe or a map of the Earth, you will see that oceans cover most of the planet. In fact, oceans cover almost 3/4 of Earth's surface!

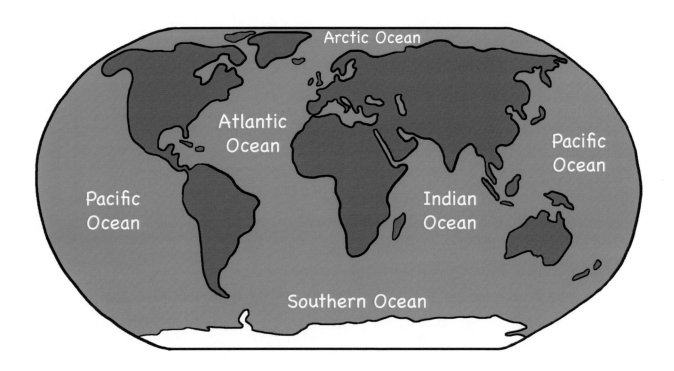

Oceans are very important for life on Earth. In addition to providing evaporated water that can fall as rain or snow, oceans absorb heat from the Sun. Ocean water is constantly moving around the Earth and carries this heat around the globe, keeping Earth from getting too cold.

The oceans gradually release into the atmosphere the heat they got from the Sun. This warms the air above the oceans. Winds blow the warmed air over the land which warms the land.

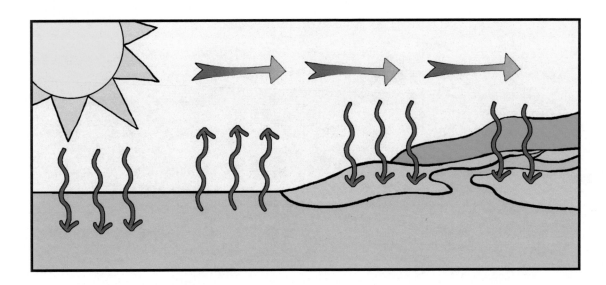

Also, air over the oceans may be cooled by the ocean water. When winds blow this cooler air over the land, the land is cooled and kept from getting too hot. The oceans have a big effect on temperature and weather on the Earth.

Even though the oceans contain so much water, the water is not good for drinking—it is much too salty to drink. The salts in the oceans come from rocks. Tiny bits of rocks are worn away from big rocks by rain, wind, and moving water. These very tiny rock bits are carried by rivers to the oceans and make the ocean water salty.

14.5 Water on the Land

Surface water is the name for the water that is on top of the land. Surface water is found in lakes, rivers, streams, swamps, and marshes. Surface water is very important for life. Animals and plants would not be able to live without water.

14.6 Water in the Ground

Water that is under the Earth's surface is called groundwater. When it rains or snows, some of the rain or snow seeps down into the ground. Plants can take up this water through their roots and use it to stay alive and grow.

There are some places in the ground that hold lots of water. Wells pump this water to the surface where it can be used for drinking and other purposes, such as farming.

14.7 Keeping Our Water Clean

The same water is used over and over on Earth. The same water will sometimes be in the oceans, sometimes on the land or in the ground, and sometimes in the atmosphere. The same water keeps changing between its liquid, solid, and gaseous forms and keeps moving around the Earth.

Without clean water, life could not exist on Earth. Since we use the same water over and over, it's important to keep it clean. But people are not always careful to keep our water clean. They throw trash and chemicals into rivers and the oceans.

Also, smoke from factory smokestacks and exhaust from cars can dirty, or pollute, the air in the atmosphere. This pollution mixes with water vapor in the clouds and falls to Earth as rain or snow. If there is enough pollution, it can make plants and animals sick.

Geologists and other scientists are studying pollution to find out how we can change the way we do things so that we can keep our water and air clean.

14.8 Summary

○ The water part of Earth is called the hydrosphere.

○ Water exists in three forms: liquid, solid (ice and snow), and gas (water vapor).

○ Water moves around the Earth in a cycle, or series of events that repeat.

○ Oceans hold most of the Earth's water.

○ Surface water is water that is on top of the land, and groundwater is water that is under Earth's surface.

○ Clean water is important for life to exist.

Chapter 15   Plants and Animals

Geology

15.1 Introduction

Earth is not made of just rocks, water, and air. Earth has trees, frogs, butterflies, rabbits, deer, and worms. Earth is the only place we know of that has living things. The living things on Earth make up what is known as the biosphere.

The biosphere contains all living things and every place where life can exist on Earth. The biosphere includes plants, animals, and bugs, and also the land, the water on the land, the oceans, the part of the atmosphere near Earth, and even some underground places.

Different parts of the biosphere work together to help support life. For example, the soil provides the water and nutrients plants need to live. Animals use plants for food and they drink water. Birds fly in the atmosphere to catch bugs for food. Plants and animals get the carbon dioxide and oxygen they need from the atmosphere.

15.2 Cycles

Water is not the only resource on Earth that is used over and over again. There are also different elements (atoms) that are used repeatedly by living things in the biosphere.

The elements carbon and oxygen are used over and over again in a carbon-oxygen cycle. Think about the oxygen we and other animals breathe in from the atmosphere. We inhale oxygen atoms that we use to power our bodies. We breathe out carbon dioxide.

Plants use carbon atoms from carbon dioxide to make food and then release oxygen atoms back into the air where animals breathe them in again. In the carbon-oxygen cycle, both oxygen atoms and carbon atoms are used over and over.

Nitrogen also has a cycle. In the nitrogen cycle, nitrogen from the atmosphere goes into the soil where bacteria change the nitrogen into a form that plants can use. This process is called "fixing" the nitrogen. Plants absorb the "fixed" nitrogen with their roots and use it to grow. Animals eat the plants and use the nitrogen from the plants to make proteins and run the machinery inside their cells.

Without the carbon-oxygen cycle and the nitrogen cycle plants and animals would not be able to live.

15.3 The Sun

Do you know how animals get energy from the Sun? By eating plants! When sunlight shines on plants, the plants use the sunlight to make the sugars they use for their own food. When animals eat the plants, they get energy from the Sun by using the sugars that the plants made from sunlight.

15.4 Environment

An environment is everything that surrounds a living thing in the area where it lives. Water, weather, soils, plants, and animals are all part of an environment.

Scientists study how all the different parts of an environment affect each other. How much water do the plants in a particular area need to have in order to grow? Which plants will certain animals eat? Which living things exist in environments that are hot and dry? Which ones live where it is cold or wet?

Learning about different environments helps scientists understand what resources are needed for plants and animals in a specific area to grow and be healthy.

15.5 Summary

○ The biosphere is the living part of Earth and contains all the living things on Earth.

○ Different parts of the biosphere work together to help support life.

○ Living things in the biosphere use oxygen, carbon, and nitrogen atoms in cycles that repeat over and over.

○ Animals get energy from the Sun by eating plants.

○ An environment includes everything that surrounds a living thing in the area where it lives.

Chapter 16   Magnetic Earth

Geology

16.1 Introduction

Have you ever noticed how a refrigerator magnet sticks to metal things? Have you played with a compass and observed how the needle always points in the same direction? In Experiment 13 you observed how the opposite poles of two magnets will stick together and the same poles will not. All of these events occur because of magnetic forces.

16.2 Magnets Have Fields

Recall from Chapter 13 that a magnet is a particular kind of metal that can create magnetic forces. Magnetic forces allow a magnet to attract certain types of metals to it. Magnetic forces surround a magnet in what is called a magnetic field.

Recall that magnets have poles, or opposite ends. The poles in a magnet occur because the magnetic forces are going in opposite directions. The north pole of a magnet is where the magnetic field points outward, and the south pole is where the magnetic field points inward.

If you have two magnets, you can find out which of their two poles are the same and which are different. The poles that are the same will repel each other (push each other apart). The poles that are different will attract each other and stick together.

16.3 Earth Is a Magnet!

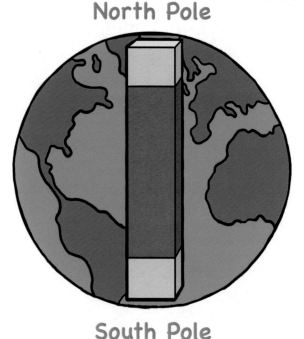

North Pole

South Pole

It's hard to believe, but Earth is a gigantic magnet with a north pole and a south pole! You might think about Earth being a magnet by imagining a huge bar magnet going through the center of Earth from the top to the bottom. The top end of Earth's magnet is at the North Pole and the bottom end is at the South Pole.

The outer part of Earth's core is made of molten iron and nickel. Scientists think this molten part of Earth's core swirls around, creating a magnetic force. This magnetic force surrounds the Earth in a magnetic field.

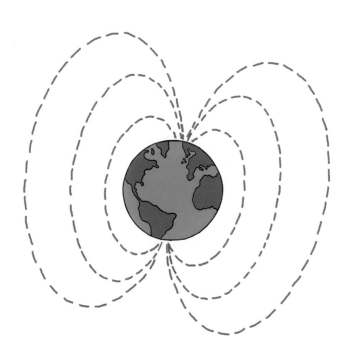

THIS COMPASS WAS RIGHT!

NORTH POLE

Did you know that you can use Earth's magnetic field to find your way out of the woods? When you use a compass, the magnetic needle in the compass is attracted to the Earth's North Pole, so the needle always points to the north.

16.4 Earth's Magnetic Field in Space

Earth's magnetic field extends into space and is affected by heat and light energy sent out by the Sun. This energy is called solar wind. Earth's magnetosphere is formed when the solar wind hits the magnetic field.

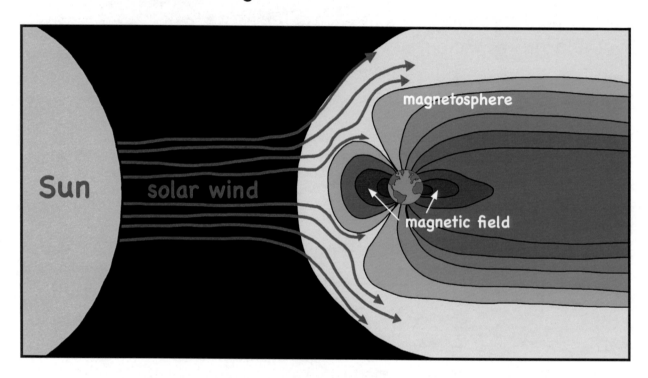

Plants and animals need heat and light energy from the Sun in order to live and grow. But too much of this energy would be harmful to life. The magnetosphere protects life on Earth by letting just enough energy get through. The excess energy is stopped by the magnetosphere. This extra energy then slides around the magnetosphere and continues on into space.

16.5 Summary

○ A magnet is a certain type of metal that can attract certain other metals.

○ Magnetic force allows a magnet to attract other metals.

○ A magnet has opposite poles called the north pole and the south pole.

○ Earth is like a giant magnet.

○ The magnetosphere contains Earth's magnetic field and protects Earth from getting too much energy from the Sun.

Chapter 17 Working Together

Geology

17.1 Introduction

Did you know that all the parts of Earth depend on each other? Without the atmosphere, plants and animals in the biosphere could not get the oxygen and carbon dioxide they need to live, and there would be no rain to bring them water. Without water from the hydrosphere, the cells that make up plants and animals could not produce oxygen

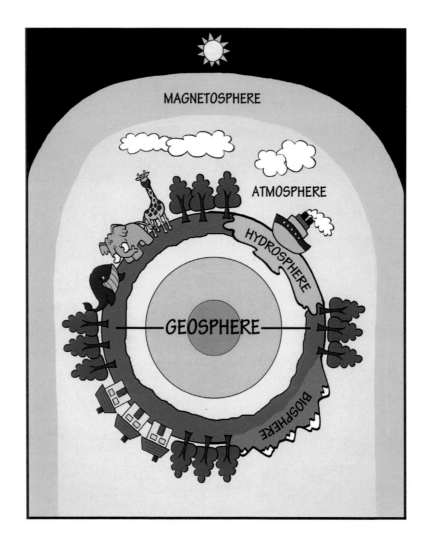

and carbon dioxide to go into the atmosphere. Without the swirling of iron and nickel in Earth's geosphere, there would be no magnetosphere. Without the magnetosphere, plants and animals would get too much energy from the Sun and would die.

17.2 Earth as a Whole

All of the parts of Earth fit together in just the right way, like a big puzzle. In order for Earth to function as a whole, it needs all of the pieces to be in place.

The geosphere, biosphere, hydrosphere, atmosphere, and magnetosphere all work together to make up what we know as Earth. All of the parts of Earth depend on each other. In order to grow and live, you depend on members of your family and members of your community. In a similar way, all the parts of Earth depend on each other to keep Earth working. If you were to take away any one part, Earth as we know it wouldn't exist.

17.3 Earth in Balance

Earth's parts are in balance with each other. There is enough liquid water and water vapor for rivers, oceans, clouds, and rain. There are enough plants to produce oxygen for animals and enough animals to make carbon dioxide for plants. There is enough of the Sun's energy for plants to grow and a strong enough magnetosphere to block excess energy from the Sun.

However, it is possible to throw Earth off balance. If too much carbon dioxide were in the atmosphere and there weren't enough plants to change it to oxygen, the Earth's climate would change. As a result, the Earth would become too warm or too cold. If too much liquid water were stored as ice, there would be less water in the oceans and they might not be able to support life. If too much ice melted, weather patterns could change, making some parts of Earth too wet and some parts too dry, some parts too hot and some parts too cold.

17.4 How Can We Help?

Keeping Earth in balance is important for life. Many of Earth's cycles can adjust to small changes, but if the changes were to get too big, Earth's cycles could begin to work differently from the way they do now. Scientists don't understand everything about how Earth's cycles work, and they don't know everything about how to keep Earth in balance.

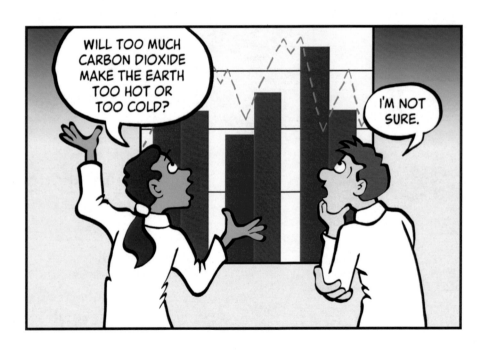

Humans can both help and hurt Earth's balance. For example, humans make some chemicals that can create problems for plants and animals. If too many chemicals are in the environment, plants and animals can die. But if humans clean up the harmful chemicals and replace them with ones that are not harmful to living things, the plants and animals will have a better chance of staying healthy.

Humans also use products, such as plastics, that can create problems when a lot of them get into the oceans. Scientists are trying to discover how to make materials that could be used like plastics but would be changed into harmless substances after being used. This would be a great step toward keeping the oceans clean.

Scientists are working on many new ideas that could help keep our planet healthy and in balance. Maybe you will come up with the next great idea!

17.5 Summary

○ All of Earth's parts work together.

○ The atmosphere, biosphere, hydrosphere, geosphere, and magnetosphere all depend on each other.

○ Earth stays in balance naturally.

○ Human activity can change Earth's balance.

Chapter 18 Galaxies

Astronomy

YOU ARE
HERE

HOAG'S
OBJECT

GALAXY
CENTER

18.1 Introduction

You probably know that Earth is the third planet from the Sun and that there are seven other planets that orbit the Sun. You know that Earth "lives" in a "neighborhood" called a solar system. But where does our solar system "live"?

Solar systems, like ours, exist in a larger collection of stars, planets, and moons called a galaxy. Galaxies are like big cities, with several hundred or even thousands of solar "neighborhoods."

18.2 Galaxies Are Like Cities in Space

On a clear night, far away from city lights, you can often see a band of stars and light across the night sky. These stars are in the Milky Way Galaxy, the galaxy where we live. The Milky Way Galaxy is like a big city of stars, planets, dust, and other objects.

In a city, there are different neighborhoods, parks, shopping areas, and roads. All of these are grouped together to make a city. A city is organized in a certain way and has a certain shape that depends on how the neighborhoods, parks, shops and roads are put together.

The same is true of a galaxy. Like a big city, a galaxy holds all of the stars, planets, and other objects together in a particular shape. We will learn about the different shapes of galaxies in Chapter 20.

18.3 How Many Galaxies?

Scientists don't have a way to actually count the number of galaxies in the universe, but some astronomers estimate that there may be at least 170 billion galaxies. Other astronomers think there may be a trillion galaxies or even more! That's a lot of galaxies!! As our space telescopes become more and more powerful, astronomers think they will be able to see more and more galaxies.

It's hard to imagine just how many galaxies 170 billion would be. If you pretend that one galaxy is the size of a marble and if you had 170 billion

marbles and lined them up end-to-end, you would need to wrap them around the Earth almost 30 times! That's a LOT of marbles!

18.4 What Is a Galaxy Made Of?

Galaxies are made of gases, stars, planets, dust, and other objects. Some scientists believe that most galaxies have a black hole at the center with the stars and other objects revolving around the black hole. This theory holds that within galaxies new stars are formed from the clouds of gas and dust and that most galaxies have enough gas and dust to form billions of new stars.

Galaxies are thought to begin as small clumps of stars and then grow as new stars are made. Scientists also believe that stars can get old and can explode at the end of their life.

18.5 Other Stuff About Galaxies

Astronomers have discovered some very small galaxies that contain only about 1000 stars. They have also discovered gigantic galaxies that are 50 times the size of the Milky Way! The Milky Way is considered to be an average size galaxy.

Another interesting fact is that everything in space is in motion. For most galaxies, the objects within the galaxy revolve around the center of the galaxy. The galaxies themselves are moving through space, and scientists believe that sometimes galaxies will run into each other and join to form a new galaxy.

18.6 Summary

- A galaxy is like a big city made of lots of stars, planets, dust, and other objects.

- We live in the Milky Way Galaxy.

- There are many billions of galaxies in the universe.

- Galaxies come in many different sizes, from very small to gigantic, with the Milky Way being an average size galaxy.

Chapter 19 Our Galaxy The Milky Way

Astronomy

PERSEUS ARM

SAGITTARIUS ARM

NORMA ARM

SCUTUM–CENTAURUS ARM

19.1 Introduction

In this chapter we will take a closer look at our galaxy, the Milky Way Galaxy. Scientists estimate that there are many billions of stars in the Milky Way Galaxy and that many of these stars may have planets orbiting them.

19.2 Our Galaxy

Think about the city you live in or one you have visited. Can you see the whole city from your house? Is it easy to tell the shape of your city from where you live?

No! Because you are one small person in a big city, you can't tell what your city looks like from your house. You would need to see your city from a different place, like an airplane or spaceship, to see what shape it has.

In the same way, it is difficult for astronomers to see the Milky Way Galaxy. No one has taken a picture of the Milky Way Galaxy because it is too big and we can't fly far enough away to see the whole galaxy. However, astronomers can guess what the Milky Way Galaxy looks like by observing other galaxies.

Is our galaxy round or flat? Is our galaxy large or small? Does our galaxy have a fixed center, like an orange, or does it move like Jell-O?

Even though we've never seen our galaxy from the outside, modern astronomers think that the Milky Way is shaped like a pinwheel. Just like a pinwheel, our galaxy has spiraling arms and a bulge in the center.

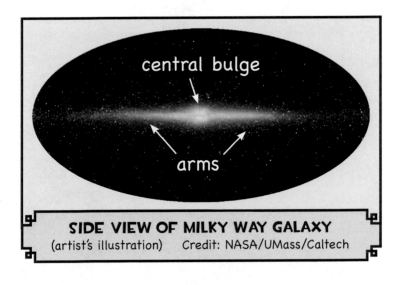

SIDE VIEW OF MILKY WAY GALAXY
(artist's illustration)   Credit: NASA/UMass/Caltech

This central bulge is a dense ball of stars. The arms of our galaxy are flatter at the edges than the center. Most of the stars in our galaxy are in the center, with fewer stars on the edges.

The Milky Way has two major arms, which are called the Scutum-Centaurus Arm and the Perseus Arm, and two minor arms, called the Norma Arm and the Sagittarius Arm. These arms spread out from the center, creating a spiral galaxy that looks like a pinwheel.

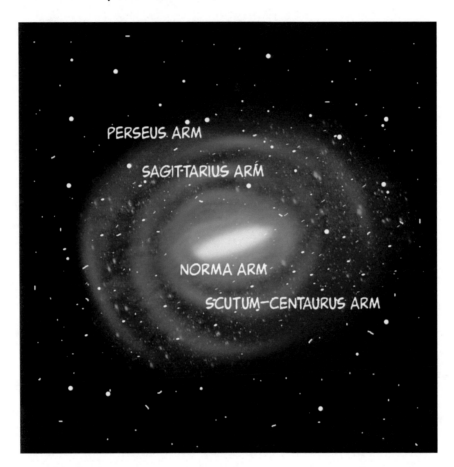

19.3 Where Are We?

Our Sun and solar system are located on a partial arm called the Orion Arm. The Orion Arm is between the Sagittarius and Perseus arms. Scientists think that our solar system may be about halfway between the center and the outer edge of the Milky Way Galaxy, but this is still uncertain.

We happen to live at just the right place in our galaxy. If our solar system were too far from the center of the galaxy, a planet like Earth might not be able to form. If our solar system were too close to the center, there might be too many stars creating too much radiation and gravity for life to form. As it turns out, we live in the right place in our galaxy for life to exist—not too close to the center and not too far away!

19.4 Earth Moves

The Milky Way Galaxy, like other galaxies, is in motion. It is thought that the objects in the Milky Way Galaxy revolve around a black hole at the center and that the entire Milky Way Galaxy moves through space. We can see that Earth, too, is constantly in motion. The Earth spins on its axis, revolves around the Sun, travels with our solar system around the center of the Milky Way Galaxy, and moves through space with the entire galaxy.

19.5 Summary

- Astronomers think the Milky Way Galaxy is shaped like a pinwheel.

- The shape of the Milky Way Galaxy is called a spiral galaxy.

- Our Earth is in the right spot in our galaxy for life to exist.

- Earth moves through space in several ways—spinning on its axis, revolving around the Sun, traveling with our solar system around the center of the Milky Way Galaxy, and moving through space with the entire Milky Way Galaxy.

Chapter 20 Beyond Our Galaxy

Astronomy

20.1 Introduction

In Chapter 19 we looked at our galaxy, the Milky Way Galaxy. We saw that our galaxy is called a spiral galaxy and that it has a central bulge and has arms that extend outward from the center like a pinwheel.

In this chapter we will take a look at other types of galaxies. From Earth, astronomers have been able to view thousands of different galaxies. Some of these galaxies look like ours, with a central bulge and spiral arms. But some galaxies look very different from our galaxy and have unusual features.

20.2 More Spiral Galaxies

There are many galaxies like ours. Spiral galaxies are fairly common in the universe. However, even spiral galaxies look different from one another.

In some spiral galaxies the central bulge is very large. This type of galaxy is called an Sa galaxy. Other spiral galaxies have a central bulge that is smaller. This type of galaxy is called an Sc galaxy. What the arms look like is also taken into consideration when classifying spiral galaxies.

Some spiral galaxies have a bar-shaped cluster of stars in the center. This type of spiral galaxy is called a barred spiral galaxy, or an SB galaxy. Many astronomers think the Milky Way Galaxy might be a barred spiral galaxy rather than a regular spiral galaxy.

20.3 Other Types of Galaxies

Astronomers have seen other types of galaxies that are different from spiral and barred spiral galaxies.

An elliptical galaxy is a type of galaxy that can look like one huge star but is really a group of tightly packed stars. Elliptical galaxies can vary greatly in size. Some elliptical galaxies are small and are called dwarf elliptical galaxies.

There are other elliptical galaxies that are very large and are made of trillions of stars. Elliptical galaxies are round or elliptical in shape and don't have any special features.

Irregular galaxies don't look like spiral galaxies or elliptical galaxies. They don't really fit into any other category of galaxies and have a variety of odd features.

Some irregular galaxies have a large bulge of stars off to one side with a ring of stars surrounding it. Other irregular galaxies are dumbbell or butterfly shaped.

It's hard to know how some of these galaxies got their shape. Some astronomers think it might be possible that some of these irregular galaxies have such odd shapes because they are galaxies that have bumped into each other. But other astronomers think that the irregular shapes developed as the galaxies formed. Learning how galaxies form is an exciting area of study in astronomy.

20.4 The Local Group of Galaxies

Just as planets exist together around a star to form a solar system and solar systems exist together to form galaxies, astronomers have discovered that galaxies exist together to form large groups. The Milky Way Galaxy is actually part of a large group of galaxies that are close together. Astronomers call

this group of galaxies the Local Group. It's uncertain how many galaxies are in the Local Group because astronomers keep discovering new galaxies that are close to ours. Some astronomers now think there are over 50 galaxies in the Local Group.

The Milky Way is one of the three largest galaxies in the Local Group. Andromeda, our nearest neighboring galaxy, is the biggest galaxy in the Local Group. And Triangulum is the third largest galaxy in the Local Group, with the Milky Way being the second largest.

20.5 Summary

● The three types of galaxies are spiral, elliptical, and irregular.

● Spiral galaxies can have large or small central bulges and can have a bar-shaped cluster of stars in the center.

● An elliptical galaxy can look like one huge star but can contain billions or trillions of stars.

● Galaxies exist together to form large groups.

Chapter 21 Other Stuff in Space

Astronomy

LET'S GO CHECK OUT THE CRAB NEBULA

OKAY!

21.1 Introduction

In Chapter 20 we looked at different types of galaxies. We saw that some galaxies are like ours, with a central bulge and spiral arms. We also saw that some galaxies are irregular or elliptical.

What other things exist in the universe besides stars, planets, and galaxies?

21.2 Comets and Asteroids

Comets and asteroids are found throughout the universe. Comets are large chunks of rock and ice that fly though space at great speeds. When a comet gets near a sun, the heat will make the ice in the comet begin to vaporize, creating a beautiful tail. When ice vaporizes, it changes to a gas without becoming liquid water first. You can notice the results of

vaporization if you leave ice cubes in the freezer for a long time. They get smaller as the ice vaporizes.

Asteroids are made of rock and have irregular shapes. The Asteroid Belt is a ring of asteroids that travel around the Sun between the orbits of Mars and Jupiter. Many other asteroids exist outside the Asteroid Belt. Like comets, asteroids move through space at very high speeds. Most asteroids are small, and sometimes they collide with one another! As a result asteroids are often covered with small craters that are caused by these impacts.

Once in a while a comet or an asteroid will come close enough to Earth to hit it. If an asteroid hits Earth, it is called a meteorite. Although asteroids occasionally hit Earth, most of the time they burn up in our atmosphere before they reach the ground. If you have ever seen a shooting star, it is really an asteroid burning up as it flies through Earth's atmosphere.

21.3 Exploding Stars

Stars do not stay the same size or generate the same energy forever. Stars actually have a birth and a death. When a star is born, it is able to generate light and heat energy for a very long period of time. However, at some point it runs out of energy and dies. When the star begins to run out of energy, it gets very large, burning brighter and brighter. Astronomers call this type of star a red giant. Once the red giant star uses up all its energy, it shrinks into a small white dwarf star.

Sometimes stars actually explode. A supernova is a star that is exploding. When a supernova star explodes, it becomes very large and bright, expanding many millions of miles into the surrounding area.

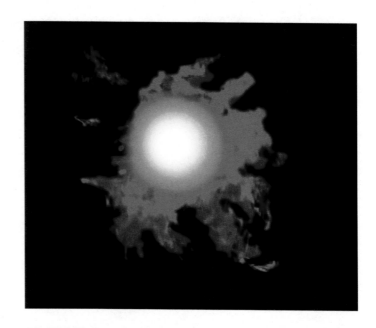

21.4 Collapsed Stars

What happens to the exploding star once it has finished exploding? Where does it go? Does it turn into nothing, or does it become something else?

Many astronomers think that after a big star explodes, it may collapse and form a black hole. A black hole is an odd feature in the universe. It is difficult to see a black hole because it doesn't allow any radio waves or light waves to bounce back from it. Because of this, it just looks like there is a dark, black hole in the middle of space.

21.5 Nebulae

By using the Hubble Space Telescope, astronomers can explore the universe in ways never before possible. The Hubble Space Telescope has given us some very beautiful pictures of stars, planets, asteroids, and galaxies.

Some of the most beautiful images captured by the Hubble Space Telescope are nebulae. Nebulae are clouds of gas, dust, and particles. The gas, dust, and particles swirl in space to create amazing celestial sculptures and cosmic art. Today several thousand nebulae have been photographed with the Hubble Space Telescope.

However, we have only just begun imaging, understanding, and discovering the stars, planets, and other objects that exist in space. Future generations of astronomers have a whole universe to discover and explore!

21.6 Summary

- Comets are objects in space that are made of rock and ice.

- Asteroids are objects in space that are made of rock and have irregular shapes. If an asteroid makes contact with Earth, it is called a meteorite.

- A star that explodes is called a supernova.

- Black holes are thought to be collapsed supernova stars.

- Nebulae are clouds of gas, dust, and particles.

Chapter 22 Putting It All Together

Conclusion

22.1 Using Science

In this book you learned about some of the different ways that modern cultures use science and how science can make our lives better and easier.

Chemistry helps scientists understand mixtures, and they can then apply this understanding to make products we use every day. Chemistry has also increased our understanding of the molecules in our body and how they make our body function.

Biology helps scientists understand living things, such as plants. By understanding plants, the parts of plants, and how plants grow, people can make improvements to plants, grow plants that are healthier, and grow greater quantities of food. Studying plants also helps us find ways to keep the environment clean and healthy.

Physics helps scientists understand the energy of atoms and molecules. Learning about the energy of atoms and molecules helps us understand chemical reactions and electricity. Knowing about chemical reactions and electricity has allowed us to power modern tools and technologies.

Geology and astronomy help scientists learn about Earth's systems and how Earth fits in the universe. Knowledge of Earth's systems and how Earth fits into the larger universe helps us develop new tools and technologies for the future.

22.2 Sharing Knowledge

Scientific discoveries and new inventions have a better chance of happening when scientists from different disciplines (areas of study) share their work.

The light bulb is an invention that happened because physicists and chemists shared their knowledge about materials and electrons.

New medicines that treat diseases are created when biologists and chemists share knowledge about how the body works and how chemicals interact with it.

Understanding earthquakes is possible because physicists and geologists have shared knowledge about force, energy, waves, and rocks. And much of what we know about the stars, planets, and galaxies was discovered because astronomers have learned from physicists and chemists.

22.3 Looking Towards the Future

Today it is very easy for scientists from all over the world and from many different disciplines to share information. Modern methods of communication, including cell phones and the internet, have made it possible to share more information more quickly than ever before.

As more people learn about science and as more scientists share their knowledge, discoveries can occur at a faster pace. Imagine what the future might look like if many of the problems we have today could be solved. Imagine what it would be like to travel on a train at the speed of sound, or convert all the trash in the world to potting soil, or deliver clean drinking water to every living being on the planet!

Solving tomorrow's problems starts with sharing information and using science to invent new technologies, discover new cures, and better understand the world around us.

22.4 Summary

● Chemistry, biology, physics, geology, and astronomy all contribute to our understanding of the world around us.

● Scientists from different disciplines share information.

● Sharing information and using science can help solve tomorrow's problems.

More REAL SCIENCE-4-KIDS Books
by Rebecca W. Keller, PhD

Focus Series unit study program — each title has a Student Textbook with accompanying Laboratory Workbook, Teacher's Manual, Study Folder, Quizzes, and Recorded Lectures

Focus On Elementary Chemistry
Focus On Elementary Biology
Focus On Elementary Physics
Focus On Elementary Geology
Focus On Elementary Astronomy

Focus On Middle School Chemistry
Focus On Middle School Biology
Focus On Middle School Physics
Focus On Middle School Geology
Focus On Middle School Astronomy

Focus On High School Chemistry

Building Blocks Series year-long study program — each Student Textbook has accompanying Laboratory Notebook, Teacher's Manual, Lesson Plan, and Quizzes

Exploring the Building Blocks of Science Book K (Activity Book)
Exploring the Building Blocks of Science Book 1
Exploring the Building Blocks of Science Book 2
Exploring the Building Blocks of Science Book 3
Exploring the Building Blocks of Science Book 4
Exploring the Building Blocks of Science Book 5
Exploring the Building Blocks of Science Book 6
Exploring the Building Blocks of Science Book 7
Exploring the Building Blocks of Science Book 8

Super Simple Science Experiments Series

21 Super Simple Chemistry Experiments
21 Super Simple Biology Experiments
21 Super Simple Physics Experiments
21 Super Simple Geology Experiments
21 Super Simple Astronomy Experiments

Kogs-4-Kids Series interdisciplinary workbooks that connect science to other areas of study

Physics Connects to Language
Biology Connects to Language
Chemistry Connects to Language
Geology Connects to Language
Astronomy Connects to Language

Note: A few titles may still be in production.

Gravitas Publications Inc.

www.realscience4kids.com